FAIRE OF LITERATURE CARTOONS, PHILOSOPHY, ASTRONOMY, PSYCHOLOGY, QUANTUM FIELD THEORY IN TOPOLOGY / CLASSICAL MECHANICS, QUANTUM GRAVITY, M THEORY / CLASSICAL PHYSICS
(a true vacuum theory)

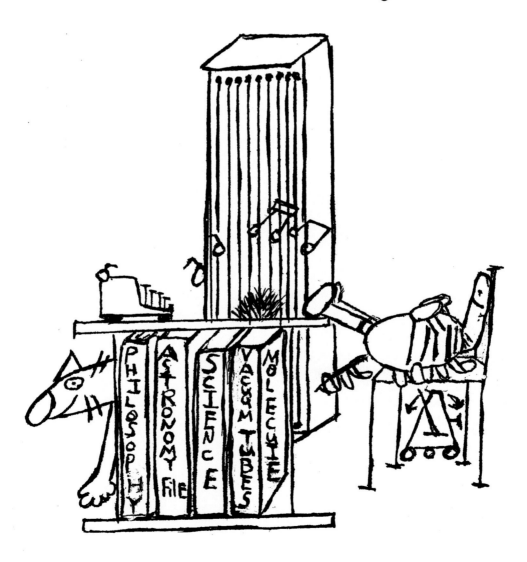

To order additional copies of this book, contact:
Xlibris
844-714-8691
www.Xlibris.com
Orders@Xlibris.com

ISBN: 978-1-4134-1830-9 (sc)

Library of Congress Control Number: 2003094338

Print information available on the last page

Rev. date: 01/11/2024

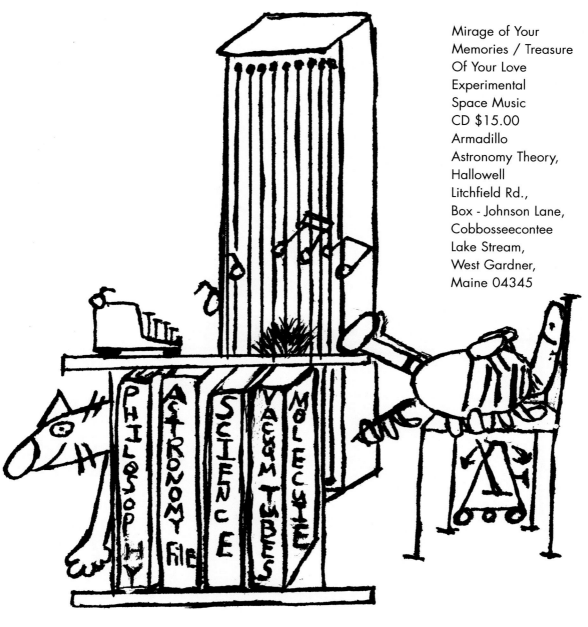

Mirage of Your
Memories / Treasure
Of Your Love
Experimental
Space Music
CD $15.00
Armadillo
Astronomy Theory,
Hallowell
Litchfield Rd.,
Box - Johnson Lane,
Cobbosseecontee
Lake Stream,
West Gardner,
Maine 04345

criticism

Where Carly Simon reviewed on T.V.
(of ingenuity with a sign of unity) from
the song "Mirage of Your Memory"

'Free Writing Style'
"To overcome writers block"

'Family dog making his own music.' In his timing of walking back and forth
past the 9 string Aeolian Harp.

'Criticism'

Criticism from the European Radio Stations, The words of the songs, Treasure
of Your Love, Mirage of Your Memory, "Are very beautiful".

Philosophy of Astronomy

Philosophy, like Astronomy; It's here, then it's gone!
where my first noted memory of remembering at 3 yrs of age.
Was a pink Cadillac car driven into the dirt drive.
Shown a trunk full of winning gambling money.
With a suit case full of money for small change.
Then the money and car all lost in a bet the following day.

Observable realm of changing particulars known as realm of the heavens fairy
In this period of time of a Nations literature of a fairy, likeness of ancient Greek literature,
"If there's a Goddess then there's math".
The illusions of white cloud molecular of mirror likeness
Can be imaginary calculated of the m/molecular weight formula unit of the "flat plane"
from the mirror planes of the "plane".

"How To Become a Genius In Literacy"

Where at 4 yrs. Of age finding my genius. Use to spend a whole Saturday in front of the huge book case.

Looking at the technical drawings and reading & pronouncing the words correctly in your mind.

Where the book case was before the 45 degree "plane" plateau of phenomena of the observable realm of changing particulars of illusion at the top of the mirror staircase, the illusion theory study that was a written contribution exploration of LYRA Star Constellation.

"Legendary neighborhood stray cat"

Aristotle's Psychology of Animals

"Field Philosophy on animals" "Friendship!"

"Animal Philosophy", on the Egg Farm. In the year of my literacy writing of the space exploration theory of LYRA star constellation. My field books; Armadillo Seven Plate Astronomy Cartoons & Exploration Of The Heavens. "Theory" at LYRA. Astronomy: Periodic Motion of a vibrating string, Natural Philosophy: (study of animals) 7 Plate Armadillo, Moral Philosophy: (ethical) observable realm of changing particulars with physics, Metaphysical Philosophy: (abstruse) flight of the arrow "Straight Line" & geometrical "plane" 'Field of Error Phenomenon'. Credit of the exploration of LYRA star constellation & Astronomy File. Credit of my file & plaque on the Air & Space Exploration Wall Of Honor, of the National Air And Space Museum Udvar-Hazy Center, Washington Dulles International Airport, D.C.
(astronomy file) IRAS (Infrared Astronomical Satellite), (NIVR), NASA, (SERC) of four wavebands 10, 60, 25, 12 um, discovery of a dust shell around *Vega, an early planetary system & discovered 5 new comets, & IRAS galaxies from the galactic plane 60 and 100 um wavebands, spirals, *protostars in several *molecular cloud complexes.
(Also the famous planetary nebula at M57 LYRA)

Aristotles Psychology of Animals
Part 2

'Field philosophy on animals' The Basketball "practice"
During my research of Quantum Field Theory on
Topology, & Quantum Gravity of the hidden Dimension.

"Entrepreneurship"

During the Cuba embargo period, sent the books ARMADILLO SEVEN PLATE ASTRONOMY CARTOONS & EXPLORATION OF THE HEAVENS "Theory at LYRA" / ASTRONOMY: Periodic motion of a vibrating string NATURAL PHILOSOPHY: (study of animals) 7 plate Armadillo MORAL PHILOSOPHY: (ethical) observable realm of changing particulars with physics METAPHYSICAL PHILOSOPHY: (abstruse) flight of the arrow "STRAIGHT LINE" & geometrical "plane" 'FIELD OF ERROR PHENOMENON' & Armadillo clothe blue print to place on the market.

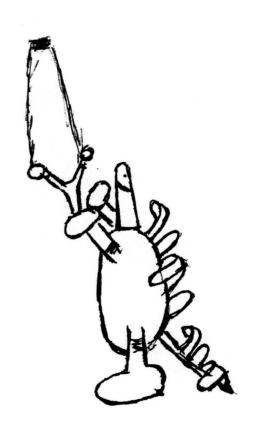

"The Day Dream"

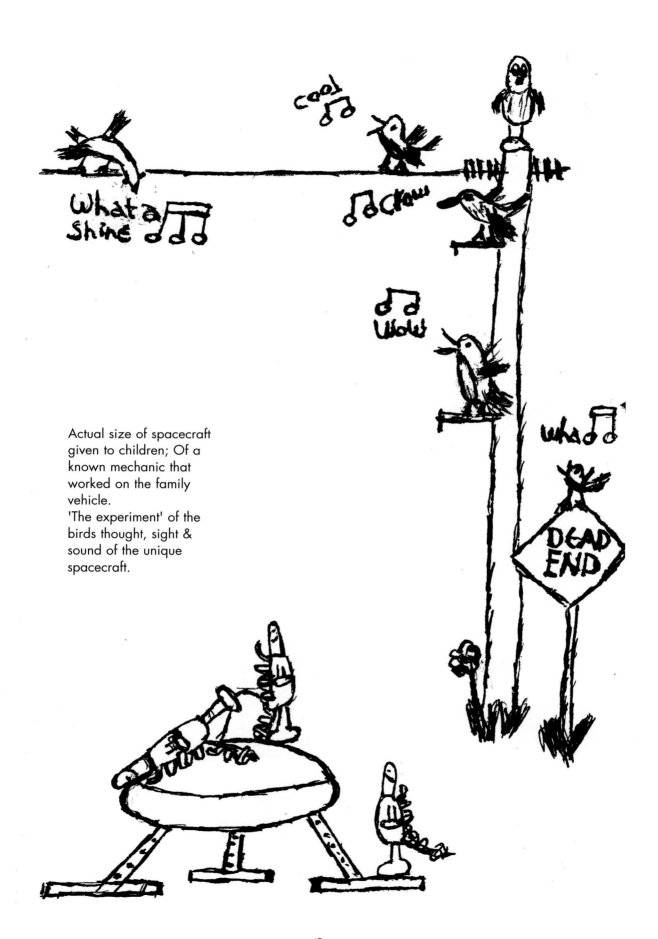

Actual size of spacecraft given to children; Of a known mechanic that worked on the family vehicle.
'The experiment' of the birds thought, sight & sound of the unique spacecraft.

Crossing A Path in Psychology "Part 2"
(Contribution in Psychology)

Military Deeply Psychology Of Hypnosis
(Computer CD Video Game Kit In Literacy)

(Where the white outline of the window
is an Optical Illusion in several views.)

origin Psychology of Hypnosis

Where at 4 yrs. Of age, around midnight at the time, workers of the lipstick container factory where getting out of work.

Where the workers cars were traveling up the hill and the car lights were reflecting onto the bedroom window, then onto the bedroom wall about 100 times.

Where I couldn't get to sleep and counted the car lights that beamed the window onto the wall. When I fell to sleep, my head felt numb and drifting deeply very fast into the darkest of outerspace and into a nightmare.

Where the following Saturday I turned on the radio to the Psychologist Station.

Where there was an ad the Military had to hire Psychologist to make deeply Psychological video games.

Draft of the Deeply Psychology of Hypnosis

(Film on DV Cinema Camcorder adapted to CD Computer, "of the computer CD Video game".) Scene of game.

Spacecrafts start at the 45 degree "plane" with 100 car lights flashing over the "plane". Where the spacecrafts try to take off in between the car lights out into outerspace towards LYRA star constellation.

Along the way to LYRA, avoid being hit by meteorites, comets, asteroids, star gas flares, and space junk. If the spacecraft gets hit will spin out into dark space, and will get lost, and get Hypnotic from passing stars.

Then back to earth of the 45 degree "plane" where again you have to pass through 100 car lights without being hit and land on the science 45 degree "plane".

If there is only one spacecraft that lands on the 45 degree "plane" wins, only if the opponent spacecraft gets lost in deep space.

(Music draft) Lead in music score, use an English bike rim replaced with thin gut music strings for experimental space music.

Rhythm in music score, use the ancient roman 80 string instrument.

Bass in music score, use the 9 string Aeolian Harp.

(Making of spacecraft models) Use clay to make spacecraft models then brush on rock hard water putty, and circus design paint the spacecraft models.

Start

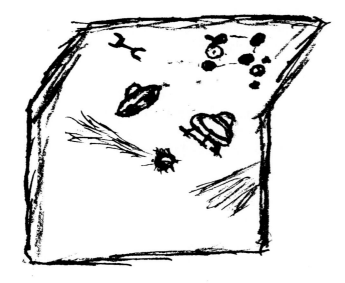

LYRA

Finish

Renaissance in NATURAL PHILOSOPHY (study of animals) observable dog:

MORAL PHILOSOPHY (ethical) observable classical mechanics in Topology phenomenon of a crystalline structure liquid drop model staircase invaluable identification of a FLAT MANIFOLD of a "plane" with the leap & spin and beauty that extends into observable Quantum Gravity from a true vacuum:

METAPHYSICAL PHILOSOPHY (Abstruse) Continuous 'FIELD OF ERROR PHENOMENON' Quantum Gravity of the Hidden dimensions of a true vacuum theory to experiment:

On a moonlit December frigid night after a snow storm. I was sitting in my snowed in mid-engine sports car for 8 hours. Nearby an airport next to an observable dog path. Where the neighborhoods huge dog stopped and placed his nose onto the car window for a few seconds then continued on his travel of backyards.

The dogs nose made a mode & 45 degree angle of incidence on the car window that was like a microscope. The dogs nose molecules known in Explorations in Quantum Physics of Handedness of the natural molecular chirality in optical activity of mirror-Image molecules chiral.

Where the molecules dripped out A CONE called the "point" of the mode. With rate of phenomenon momentum of spin and spin at angle after splitting onto the "plane" of snowflake well: Smaller particles spin & leaped toward left, larger particles spin, leaped towards right in the true vacuum phenomenon theory to experiment.

Field of the "plane" of the well in the Vector field. The field of the well in the Poincare duality theorem. "The Flat Plane" of the well field of flat tangent space.

Where the molecule would spin & leap down the stair case of the (orange color of molecules) snowflake wells where there was one delay of the spin & leap inside the well for 5 seconds towards the final spin & leap/Field of Poincare duality theorem. Superstring theory of Quantum Gravity a genuine quantum field theory.

Where at the phenomenon 45 degree of the dog nose mode of the vacuum model staircase. There were 5 beams coming from the snow flake wells. Where one beam of the middle outstood and you could study the quantum gravity of the Hidden dimensions.

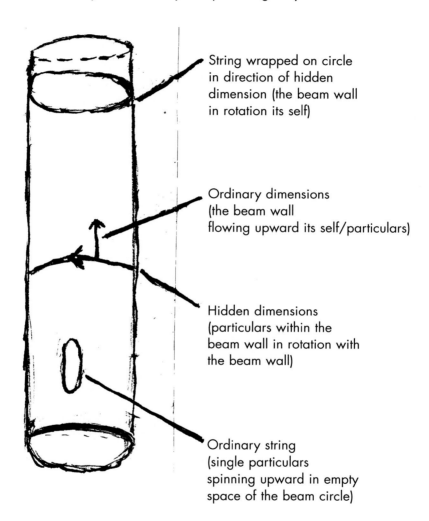

String wrapped on circle in direction of hidden dimension (the beam wall in rotation its self)

Ordinary dimensions (the beam wall flowing upward its self/particulars)

Hidden dimensions (particulars within the beam wall in rotation with the beam wall)

Ordinary string (single particulars spinning upward in empty space of the beam circle)

Of the older Shoenflies point-group notation of Theoretical chemistry & Solid state physics of Crystallography/based on rotation axes, mirror planes, of rotation-reflexion/alternating axes. M/molecular weight formula unit of the "Flat plane".

Also the white cloud moleculars sphere unit cells of spin within the beam walls of the snow flake wells.

Quantum Field Theory in Topology/Classical Mechanics

QUANTUM GRAVITY, M THEORY/CLASSICAL PHYSICS (a true vacuum theory)

('THE THIRD DIMENSION' nearby an airport of the study of the Quantum field theory)

(or MAN'S BEST FRIEND)

(For further study, sent the true vacuum theory of the Quantum Field Theory in Topology & Quantum Gravity M Theory of the well & beam to Bohr Institute for Theoretical Physics, University of Copenhagen, Denmark)

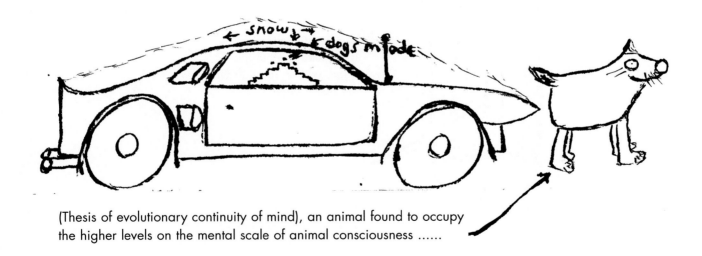

(Thesis of evolutionary continuity of mind), an animal found to occupy the higher levels on the mental scale of animal consciousness ......

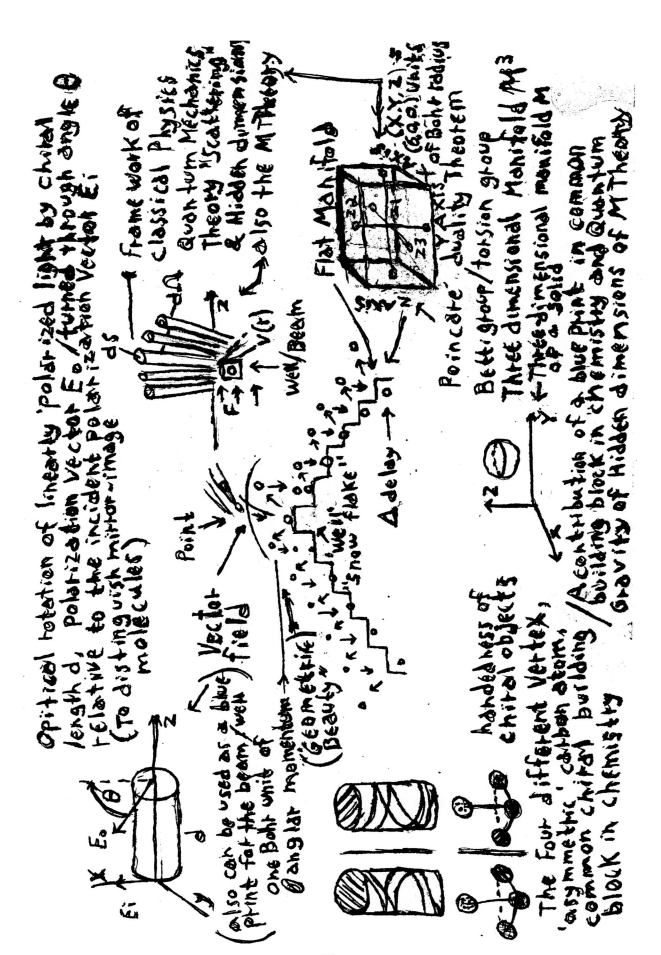

Optical rotation of linearly polarized light by chiral length d, Polarization Vector E_o /turned through angle θ relative to the incident Polarization Vector E_i (To distinguish mirror-image molecules)

Frame work of classical Physics

Quantum Mechanics Theory "Scattering" & Hidden dimensions →also the M Theory

ds

z

$V(t)$

E →

Well/Beam

Flat Manifold

axis (X,Y,Z) → (690) units of Bohr radius

Y AXIS

Z

$Z1$ $Z2$ $Z3$

Poincaré duality theorem

Betti group/torsion group

Three dimensional Manifold M^3

←Three dimensional manifold M of a solid

A contribution of a blue Print in common building block in chemistry and Quantum Gravity of Hidden dimensions of M Theory

Point

Vector field

(Geometric) Beauty

angular momentum

"Well" "snow flake"

Δ delay

handedness of chiral objects

The Four different Vertex, asymmetric carbon atom, common chiral building block in chemistry

X E_o θ E_i y z d

(Also can be used as a blue) Print fat the beam/well One Bohr unit of angular momentum

(replica
of a country
western long
wool orange
scarf ...)

Making history, first draft copy of the novel format.
From the opera, Northern White Pine.
Along with the photo of the famous
Arrow "Straight Line" of Aristotle's
invisible lines: (Zeno's) The moving
arrow must at any moment be at rest,
but no sum of resting arrows will give a
moving one.

Printed in the United States
by Baker & Taylor Publisher Services